*For My Grandparents.*

*Peter & Mavis Lidyard.*

*Whose nature inspired my every aspiration...*

**Also Published By The New World Thought Disorder:**

Aspire To Inspire
Divine Design: Soul Nature
Divine Design: Spirit Buzz
Divine Design: Mind Shine
The Research Economy

# The

# Infinity

# Theorem

*-Free Energy & The Zero-Point Power Source*

*By Alan Peter Garfoot Jnr. Cert. H.E.*

The Infinity Theorem: Free Energy & The Zero-Point Power Source.

By Alan Peter Garfoot *Jnr. Cert. H.E.*

Cover Art: 'This One Is About No-One' © A.P. Garfoot 2020

ISBN: 978-1-4717-2432-9

18/04/2022

London.

# Contents:

# Abstract:

This is an technical and theoretical exploration of the potential for the high temperature superconductor ceramic Yittium-Barium-Cuprate-Oxide, in the form of ceramic disks, to exhibit gravity influencing, altering or dampening effects. The variables that are relevant are: rotational speed, vacuum tension, temperature and floatation magnet strength; with refinements build upon the original gravity shield design. The temperature is maintained by vacuums and liquid hydrogen cooled gravitoelectrical components to zero-point temperatures, which are then suspended and held in vacuums at all connective points by previously charged ferrite floatation magnets which do not require input.

This creates a zero energy loss and entropy bleed system of constant gravitonic-kinetic-electric energy transference, which contributes through the conservation of input energy and cumulative energy gain into a feedback loop of limitless exponentially increasing output energy through reinput energy accumulation. Through experimental observations of the energy dampening effects of high temperature superconductors, upon gravitonic field/particle/wave energy, I rationalise the synthesis of the theoretical paradigm of Hyperparticle Gravitation out of the current paradigms of Zero-point, Hyperspace, Superstring, Photoelectric Ether, Antigravity and Quantum Mechanical paradigms. (Kuhn, T., (1962)

Through which deducing the existence of a Gluon-Electron-Graviton energy interaction occurring, which I then use as the basis for the formulation of the Hyperparticle Gravitation Paradigm and the Infinite Energy Theorem. Thus providing the explanation for the anomalous gravity dampening effect of the Gravity Shield and providing a rough technical outline for the creation of a Gravitoelectric

Cumulative Transference Power Source: the safest, cleanest, most economical and freely abundant energy resource available to the human race.

# **Introduction:**

The gravity influencing effect which is resultant from a zero-point electro-acoustic resonance, that can be effected in the rotating superconducting ceramic disk of a Gravity Shield. Can be harmonised in the disk to a great enough threshold of effective intensity through quantum coherence, to have a hyperspatial gravity dampening or reducing effect, on all objects that are subject to that field of Gravity that the Gravity Shield is operating within, that are placed directly above it.

This article is then set to look into whether or not the transgressive Gravity-Gluon particle-wave energy exchange and dampening effect causation of the Gravity Shield, can be harnessed within a technological system, so as to become a potential renewable and clean future energy source. Harnessed by the energy accumulator through it's capacity for cumulative energy increase and storage from a constant energy input to produce an exponentially increasing energy accumulation and output from the system.

This is where the gravity reduction or 'shielding' effect can be utilised to create a constant conserved re-input of Kinetic rotational motion, effected through a partial weight loss, upon one half of a hinged and balanced codependent source of potential gravitational energy, in the form of two opposed sides of a middle pivoted rotating flywheel. A flywheel whose two halves of balanced potential Gravitonic energy, as a source of causation, are entwined as one. A decrease of relative weight and loss of potential gravitational energy through the Gravity dampening effect of the gravity shield, to either of its two sides. Creating as an equal and opposite reaction to the causation of the gravity shield an increase of relative rotational motion of the flywheel, through a shift in the balance of the rotational

8

gravitational energy in the system that is created by the influence of the Gravity Shield.

This article will explore this gravity anomaly as a possible source of Gravitonic to Kinetic energy transfer, to see if a constant source of rotational energy, can be harnessed from the Potential Gravity energy shift, in the balance of a rotational flywheel, that can be applied as a constant energy input into an electromagnetic energy generation system.

If a constant input energy can be adequately conserved within a vacuum sealed zero-loss frictionless energy transference system, of a sealed entropy isolated causation containment and energy loss nullification. Then this can be used as a basis for the creation of a self sustaining or accumulating transference energy to output energy yield where through using floatation magnets, vacuums and supercooled high temperature superconductive gravitational energy field influencing rotational components a constant input of energy into a cumulative system can be achieved through the manipulation of gravity and time.

The cumulative energy increase can be sustained to a great enough degree where the hyperspatial field quantum gravity ether energy dampening effect of the Gravity Shield can be used to create variable amounts of accumulated kinetic torque within the power system from a fixed amount of Gravitonic energy displacement. This when combined with the kinetic momentum energy re-input into the dynamic, through the feedback loop of energy conservation at work within the system, creates from a constantly diminishing vector of energy loss and entropy depletion, a cyclic self sustaining loop of time vector dependent conserved re-input energy.

This is the dynamic of a progressively self sustaining energy source, through a constant input of energy, into a successful energy accumulating power system, through the increase of conserved and accumulated kinetic energy stored within the system. The amount of energy stored within the system as torque, and the output of energy from the system through employing a cylindrical frictionless dynamo generator as electricity, is attained, through tapping as a power source, the constant energy transference within the system, from Gravitonic to Kinetic to Electrical energy.

This threshold of energy output should increase exponentially from the cumulative build-up of energy from within the system and can be calculated as an energy input to energy output accumulation footprint from the system equations. The potential energy output to the accumulator is by definition infinite, or without definable limit or end. So as to its potential available output threshold, or end to the resources sustainability as a source of electrical or kinetic power? We have, quite literally, infinite energy.

# **Theoretical Background:**

Ever since the dawn of culture and time scientists and intellectuals have wondered as to the nature of reality and the laws which govern the true nature of the universe. Aristotle of the Ancient Greek Philosophers, questioned as to the fundamental elemental nature of the material universe and in his subject founding book Physics asked why it is that fire and air rise through levity and earth and water fall towards the ground through gravity (Aristotle, 2008; Sachs, J., 1995).

Gravity itself is invisible, but the effects of gravity indicate it is the attractive force which exists as a property of all matter, as it's attraction to all other matter, in the physical universe. As a force it exists connecting through essential nature all atoms, hadrons, leptons and objects of mass to each other, regardless of distance, and which decreases and increases in effectual intensity according to the two objects location, proximity and the the inverse square law: if you double the distance between the two objects, then you half the strength of the effect of gravity between them.

All leptons, hadrons and the objects they compose which we experience as physical substance have mass relative to the size of their manifestation and their density on a chemical, atomic and quantum level. All objects as the particles of which they are made also in a field of gravity exhibit weight; their mass multiplied by the strength of the field of dominant gravitonic effect: which at sea level on Earth is 1, hence why as a ratio for mechanical calculations mass and weight are sometimes considered interchangeable statistical values.

Albert Einstein's General Theory of Relativity (Einstein, A., 1916; Einstein, A., 1922) builds upon Newton's Classical Mechanics (Newton, I., 1726) by proposing an analogy of the essential nature of space and time as to be like a fabric. A fabric which warps and flexes as a ratio field effect according to the intensity of all operating fields of gravity and all objects of mass and weight that are considered in the calculations. This folding of the curvature of the fabric of spacetime is according to the effect of an objects mass, as an effect of attractive gravitonic force, as effected through the stretching and warping of a photoelectrically conductive and gravitonically tensile multidimensionally unified hyperspatial ether.

The distortion of the temporal nature of reality, according to the warping effect that gravity has upon the fabric of spacetime and the resultant stretching effect that has upon the motion of massless photonic quanta, according to the warping and stretching of the photo-gravitonically conductive ether, through which their electromagnetic and gravitonic energy is transmitted, demonstrates their fundamentally interwoven nature. The folding of the spacetime fabric of reality is relative to the mass of the objects considered and the strength of their fields of Gravity, according to the gradients of this fabrics curvature, created through their gravitonic attraction. Through the energetic medium of a hyperspatially unified photogravitionically conductive and mailable tensile grid of zeropoint ether.

This hyperspatial grid of what Einstein called his fabric of spacetime we must remember however is deeply rooted in analogy and therefore limited in its two dimensional application as a visualisation of causation to our understanding of the dimensional essences of reality, and the energetic interactions which exist between objects according to our schema of the fundamental laws of nature. The theories of Relativity allow us to understand the manifest properties of singularities and singularity warping effects of black holes upon spacetime, but overstates that analogy, and distorts the truth as to a black holes manifest finite nature, through speculating the fabric ripping potential of the fold of a black hole's time warping event horizon.

To expand upon Einstein's two dimensional fabric of spacetime and advance a new three dimensional formulation of his theory of gravity would be to propose not a two dimensional visualisation of a fabric rather a three dimensional tensile grid of hyper-spatially unified multidimensional electro-gravitonic ether causation. In the hyper spatial visualisation of gravity as of being of multidimensional properties to its particle-wave manifest nature would extend beyond the quantum paradigm according to the simplifying unity of higher dimensions which includes in the standard model a minimum of ten from which we must expand our theories in order to fully comprehend the true nature of gravity as a multidimensional concept.

The Graviton has the meta-spatial property of the infinite range of its location as a point particle, in terms of its distance from the object of atomic mass, as of being determined as occupying a concentric spherical locality of manifestation around the atom. Rather like the multiple point locality of an electron in an electron cloud surrounds the atom, than as a single point of individuated mechanical quantum cause and effect. The meta-temporal property of moving from the atom, to the edge of time and space, then back to the atom, happens in an instant from the perspective of the lower dimensions in an instantaneous moment of perpetual manifestation, as infinite in range as it is in its divisibility.

As the graviton pulse moves through nearby matter it picks up and ejects the latent gravity charge from the Higgs Bosons of that matter as it moves through neighbouring atoms gluon fields, in reciprocation with all other gravitons, extending infinitely through all matter in the universe as it doers so. When the graviton passes through the matter of the gluon field it absorbs and releases that ejected charge causing a resulting spasm in the origin of the pulses range and direction of location as the wave travels meta-spatially and meta-temporally towards and away from the atom.

# Experimental Methodology:

To begin the real experimental side of this discussion of energy transfer, cumulative conserved increase and the age old fable of perpetual motion we should first outline some essential analogies to really set the stage for considering the problem: how do you create from a limited input energy resource a self-evolving energy dynamic. That ultimately considering the limited, balanced and finite logical nature of the universe, how do you get more energy out of a power system, than you put in?

To start with I shall borrow the following example from astrophysics; if you have an object that's weightless, out in space and I exert a force upon it and move it at a given speed, in a certain direction, then unless some other force comes to act upon it, and change its course of motion, then it will continue to move at that speed in the direction it was initially pushed. Or take for instance of the example of a winter skater on an ice rink, let's say he or she isn't very good at skating and they push themselves off from the edge and then with trepidation skate in some direction and then stops skating so as to not accelerate themselves so they do not lose balance. They will still continue to travel in the direction they initiated towards, at the speed they pushed themselves, even though they are not exerting any force to propel themselves in going in that direction any longer.

These are two examples of where a system of causation of motion exists created from a initial spark of input energy which produces a constant or prolonged period of manifest directional energy output through a waste energy reduction or negation factor operating within the defined boundaries of the system.

The principle of causation as a metaphysical epistemology and the laws of the conservation of energy in

Newton's thesis on Thermodynamics that govern the quantum entropy state of an energy systems confines depict that the total amount of energy you can harness out of a system through the fundamental nature of reality, nature & Physics cannot be greater than the total amount of energy which was initially put into the system. So to investigate this further in the case of quantum entropy, much as in all particles of matter are physical and can have their trajectories of motion, their cause and effect, upon a tangent. Any increase in the amount of energy within the system at the quantum level of the individual particles of a gaseous body of compressed air interacting, moving around and colliding, do so at a certain speed and collision rate in a pressure defined by their motion and proximity to each other.

To briefly consider, if the pressure decreases of the gas then the concentration of energy dissipates and the heat decreases as all of the particles are no longer in local kinetic causation with each other and the energy of the system moves from a state of complexity, chaos and high energy to to a state of simplicity, order and low energy; from a localised concentration of high strong energy to a weak and diffuse state of low energy. I use further this example of the nature of the organisation of particles as heat energy at the quantum level as of consisting of the interaction of individual components as part of a composite whole which is unified with itself as an intricate web of causation and causality connecting the interacting laws of nature which govern and bind the system together.

In as much as heat and pressure correlate with each other, meaning the more heat you have, when to say trying to melt a metal, or any solid body, the less pressure you need to achieve your goal, and likewise, the more pressure you have, the less heat you need to trigger a phase change of the compound. So with the example of a gas, if you compress it the amount of energy exchanged between the particles when

16

they collide, and because of the decrease in volume, the particles impact each other more frequently and through the transfer of energy at the quantum kinetic level from particle to particle the amount of energy each particle collides with increases as well as the thermal energy increases.

With heat and light their wavelength upon the electromagnetic spectrum are represented through the medium of a photoelectric ether, which conducts the electromagnetic spectrum throughout the cosmos. This ether is a multidimensional manifestation of the indivisible expansive six dimensional unity to reality formed by the implosion of the six hyper-spatial dimensions at the point of the universes manifestation, which through its operation determines the movement of photons and their fundamental properties as energy packets of a dual wave-particle nature. The contact surface of a multidimensional waveform, a composite of many wave lengths of photonic quanta, is composed of concentric ripples that are themselves interwoven into the amalgamated composite we perceive as light.

When we come to see particles as multidimensional objects woven at a fundamental level out of the six dimensional tension which is hyperspace, then the nature of their interaction makes a lot more sense. If we imagine the vibration of a particle radiating energy consider also the ether-tension of the surface skin of that particles superstring vibration as the object emits these quantum packets of interlocking energy which ripple out their vibrations out though hyperspace into the universe.

Energy can neither be created or destroyed but through the constant conjunctions and causalities at the quantum level allow it to transform from one fundamental state of coherence to another. Where each fundamental state of quantum coherence having its own individual qualities and properties of form and manifest essence makes it in its

17

own right distinct and unique; different in its fundamental nature to other energy states. According to conventional physics and classical theorists energy can neither be created nor destroyed, you cannot get more energy out of the system then you put in, it is the nature of cause and effect that when two forces interact there is a transfer of energy from its initial state into its new form after subtracting the energy expenditure occurring from the process of the nature of the causation connecting them.

Through manipulating the connecting operation of the mechanism of action uniting the two, usually in the form of a manipulated transfer process, the resultant physical change occurs, by-product is produced and/or the energy loss occurs. It is from this process of transformation, energy loss and entropy bled from any process of energy transformation, or source of power or energy production, over a sufficient period of time, ends up diminishing to a total loss of energy at a final predestined causation determined motionless rest state as the product of a self-deprecating energy loss system of entropy decoherence and decay.

But laws of nature, matter and reality interact and there is an interconnectedness between their boundaries, and where there is interaction there is a change of energy, and fundamental state; such as when heat and pressure combine deep in the planet to fuse low pressure graphite carbon into high pressure diamond crystal. The energy input we have in the form of the heat and of the pressure has not been lost in the transformation process, it has transformed into the property of the near indestructible durability and impact resistance of the diamond, having become formed of a physical solid representation and manifestation of the initial energy intensity of its formation into its current crystalline state. This is a Geological example of how energy and matter are both composed of different variations the same fundamental manifest causal essences, yet are configured at opposite ends of the mechanisms of action to connect, interact and share their momentary quantum essence in the duality of realities interlocking laws of nature, as the universe plays out according to a finite intricate web of cause and effect until its final zero energy rest state.

So how in that case can you get infinite limitless energy out of a finite electrical power input, when into it you only put a finite amount of startup energy? The trick here is this, the heart of all of the most ambitious attempts to technologically harness gravity as an energy source and realise the ancient philosophical dream of perpetual motion. To create through technological creativity and scientific innovation a self-sustaining sealed energy system of entropic isolation and of cause and effect conservation and reuptake of the initial spark of input energy into a self perpetuating feedback loop upon a timeline of indefinitely extended output efficiency.

That through creating an energy conservation system where there are no losses through the various interacting dynamics of cause and effect at all fundamental levels, the system comes to accumulate the input energy in the form of motion, upon a vector tangent of plotted energy input and output according to the total period of the devices elapsed operating time. The systems reinput factor of accumulation is maintained by the internal kinetic energy feedback loop of the progressive self-evolution of an exponentially increasing causal dynamic of cumulative energy end state reinput and output.

The science of superconductors is one area of Physics where scientists are trying to create a perpetual feedback loop, through creating in a metal alloy, a circuit of self-sustaining electrical current, in a self sustaining perpetual cycle of flow; lacking any energy losses through harnessing the quantum interactions between the nuclei of the superconductor atoms interacting with the electrons it conducts through  the overlapping electron shells of its conductive metallic bonding to a sustained level of zero electroconductive resistance. Therefore if the flow of the current that is sustained in the superconductor can be tapped for its electromagnetic energy, without reducing the effect of the sustained dynamic of that flow of energy, through harnessing the feedback loop mechanism of action, of its cyclic self sustaining electromagnetic flow through the superconductor, then the energy and materials sciences have realised the zenith of their dreams: an infinite and green source of limitless clean free energy.

And this is my fundamental point, through creating an energy output/reuptake feedback loop which does not lose any of the initial spark of input energy that initiates the systems energy magnifying dynamic of the gravity transferred into motion created from the Gravity Shield's displacing and dampening effect of the weight displacement upon the Flywheel. This is the theoretical foundation for the Hyperparticle Gravitation Paradigm and Gravitoelectric energy production. A leap of faith which takes us from self dissipating energy systems and loss of causation of classical physics through the self-sustaining energy systems of the here theorised of the temporal extension of the lifespan of an energetic lossless dynamic of feedback self-perpetuation. Through to the possibility of advancing our technological capacities to the creation through a perpetual input of energy, into a cumulative conservation initiated perpetually accumulating feedback loop based system of energy and entropy self evolution where you really can get out more energy out of the system than you initially put into it.

So on the one hand we have the theory from the field of superconductor physics, stating that you can potentially create a self-sustaining feedback loop energy dynamic, at the quantum electromagnetic level, through an energy loss factor reduction, and entropy state decoherence nullification to create a state where the perpetual reinput of a cyclic flowing electromagnetic current, has the capacity to determine a state of constant flowing energy reinput, through its field motion of perpetual lossless flow, creating a self sustaining source of power output and system of electrical energy production. Yet we also have it stated according to the classical physics of energy state transference and mechanical motion that a situation where perpetual motion, as an actual and real, not theoretical and ideal state of mechanical manifestation of technological achievement is through it's very nature and definition, impossible. Due do to the interacting forces of the manifestation of and resistance to energetic motion and transference expelled through the transformation process would eventually reduce the energy state of any system of causation to an end state of zero eventually.

So therefore in order for the output energy to be constant, the input energy has to be constant also, as otherwise through energy losses of resistance and waste caused by the fundamental interactions between states of energy and states of matter. These interactions would cause the energy transferred in the form of its threshold of output energy, and the entropic energy bleed out of the system in the process of its transference, as waste energy, cause a gradual decline in the energy output generated by the system, along a tangent of time, until all the energy is expended and expelled and the system comes to a gradual state of rest. To synthesise these points together, if we can create a self-sustaining energy system of cause and effect, no matter what the medium of manifestation, then perpetual energy is the inevitable obvious consequence of this system of causations essential nature.

To create a self-sustaining cyclic feedback loop of entropy reinput, where the maintenance of a constant state of initial manifest input energy, within the system, comes from two things: the initial energy input into the system, and the energy that has been fed back into the system through the lossless end-state output being put back into the beginning state of the accumulated input of energy to a practical and realisable degree of energy loss, or waste, through the nature of these interactions, to result in an output tangent lifetime which exceeds the threshold extended as the plotted line of perfect efficiency.

This is not the breaking of the fundamental laws of physics, through the creation of a system of perpetual delusion as to the laws nature and energy loss dynamics of motion, and process of its transference from one state to another. This is the philosophical Cartesian smashing of the whole edifice to pieces, and crushing what remains to the particulate dust of the fundamental causal constituents of which it is made, whilst proclaiming that through the creation of a theoretical synthesis, for a self evolving energy system, through considering the potential possible creation of a progressively accumulating increase of energy through the conservation of a sustained and constant flow of input energy that we can produce an abundance of energy, without definable limit, on any scale, for any purpose, in any dominant field of gravity for any desired length of time.

Even since ancient Greek times, scientists and natural philosophers of all schools of thought the world over, have argued about the harnessing of gravity as a source of constant energy, through the possibility, and inevitable impossibility, of perpetual motion. For this technical desire and scientific ambition to be achieved through a mechanical innovation, or through an exploited dynamic of interacting forces and their mechanisms of action in the fundamental laws of matter, nature and reality may through the innovations of unique novel insight actualise the intellectual humanist aspiration, to the noble and incredible ideals and endeavour, of solving the human races energy needs for its future growth, and current day worldwide energy crisis forever.

So allow me here to venture not into madness, believing through an intellectual weakness and illogical lack of applied reason in my ideological constitution in the subjective fictions of my fairytale of objectified falsehoods, through academically inquiring after, scientifically pursuing and holding reasonable faith in the impossible realisation of the dream of somehow creating out of illusion and gullible misdirection the schematisation a system of energy causation of a similarity and likeness to perpetual motion. But here I have not only done that, but I have ventured beyond its confines into the construction a new paradigm on the outermost edge of theoretical physics, where energy we can be produced, to any extent, application or degree desired, for any length of time, on any scale we can possibly imagine its need.

For these are not the hopeless pipe dreams, of a bored hapless philosopher who has succumbed to an edifice of naive maturity and the illusions of his overactive imagination. These are the profoundest realisations, of the unique insights of a creative metaphysical explorer, pursuing the expanding edge, of a scientific frontier, which is defined only by the limitations and capacities of human reason, logic, potential, creativity and imagination. So now we have outlined the essential dynamics of the fundamental causalities which would logically, according to the rules of scientific culture and the innate causal determinants which form realities laws of nature, if not create, then at least envisage the possible practical realisation of a self-sustaining or self evolving energy system.

Through creating an initially self-sustaining system of output energy reinput through the perpetual conservation of the initiating spark of the motion inducing spark of gravity displacement caused through the energy dampening effects of the Gravity Shield, as a constant fixed amount, from which an exponential increase of output torque can be effected. Then adding to this initial factored amount of energy multiplication, through making the initial spark a constant perpetual input, then through the energy systems degree of realised input conservation through the presence of a waste energy nullification feedback loop at work, cause the energy stored within the system as so to cumulatively increase the harnessed output threshold of sourced energy along its defined period of elapsed time, perpetually increasing without limit as it goes, through the build-up of internal accumulated energy stored within the system.

So as long as this degree of the constant input and progressive increase of energy fed into the system, which can be technologically manipulated, controlled and then multiplied into the end state potential gravitational energy which can be harnessed, through employing a frictionless electrical dynamo in the form of the Generator Array and Energy Uptake Array as a resource as limitless as the period of time that elapses, and field of gravitational effect that the device operates within the once theorised noble scientific and humanist dream of solving the practical condition of the energy crisis of the human race, through the capacities of this new cultural paradigmatic bud scientific theory and technological innovation, at last be finally realised.

# Technical Design Specifications:

A: System Pan Overview:

This rough blueprint slice is pan down showing the arrangement of the uptake array, generator array, floatation Magnets, Rotation Induction Flywheel & The Gravity Shield.

B: Floatation Magnet Array:

This one is a cross section of the Floatation Magnet Array, to show how multiple bar magnets can be used to suspend a further central bar magnet which rotates on its length axis, allowing for frictionless contact between moving and non-moving components reducing Axial Resistance to zero.

C: Rotational Flywheel Cross-Section:

This diagram is a cross-section through the Rotation Induction Flywheel and the Gravity Shield to show how the 5% mass reduction effect of the operating Gravity Shield can be used to induce the motion of a centrally pivoted flywheel when placed under one half of it.

D: Uptake Array Cross-Section:

This next diagram is a slice width-ways through the Generator Array and Uptake Array Faraday Cage to show how the generated torque from the flywheel is converted into electricity through having the Generator Array rotate inside the Uptake Array to generate the current.

E: Gravity Shield Cross-Section:

This final diagram is a slice across the width of a custom made Podkletnov Gravity Shield to show how the dynamics of the system can be reworked to incorporate vacuums and floatation magnets to reduce the energy bleed of lost entropy to zero creating a further Sealed Entropy System which creates a constant, self-sustaining gravity reduction effect of 5% Mass, sustained at 9000rpm.

# Conceptual Equations:

1. Weight & Momentum:

The total weight in Kilograms equals the axle beam weight in Kilograms added to the generator array weight in Kilograms added to the flywheel weight in kilograms. This is then added to the weight of a single generator array magnet in Kilograms multiplied by the number of generator array magnets there are in the system. This is equivalent to the flotation magnet strength, divided by the load bearing strength of the floatation magnet array, which will equal the number of magnets necessary, of that strength, to hold the weight of the suspended components and to achieve adequate floatation levity at either end of the axial rod.

2. Gravity/Torque Transfer:

The input energy is basically the torque of the spinning flywheel minus the revolutions per minute of the gravity shield rotating disk. Which is equivalent to the start-up energy to get the gravity effect going from the spinning superconductor disk spinning added to the gravity shield maintainable energy gives the gravity effect transfer into rotational momentum.

3.The Energy Conservation Feedback Loop:

Flywheel rotation multiplied by the rotation rate of the Flywheel multiplied by the weight of the Flywheel, gives us the Torque as rotational momentum squared. The cumulative energy increase minus the energy re-input dynamic is equivalent to the energy accumulated minus the input energy.

## 4. Energy Lost Through Entropy Bleed:

Gravity shield input energy, plus the magnet input charge energy, added to to the vacuum charging energy; gives the energy expenditure graph and this equations baseline starting point. From which all energy systems either evolve, sustains themselves or depletes through energy loss. The total loss can be charted on a graph from this baseline. Add to it the combination of the gravity shield air resistance, minus the gravity shield vacuum charging energy, plus the gravity shield maintenance energy which is added to the array axial friction minus the floatation magnet field tension. Add this to the generator array air resistance minus the generator array air vacuum tension. Multiply this by time and you can plot the net total energy loss out of the system as a vector on a graph.

## 5. Uptake Array Output Threshold:

The Torque multiplied by the number of magnets there are in the generator array multiplied by the individual uptake zone energy yield output, then multiply that by the number of uptake zones and the size of each generator magnet then finally multiplied by the strength of each individual generator magnet; gives us the energy output of the system.

## 6. Energy Transfer Dynamics:

The ferrite magnet strength equals the electromagnetic field uptake material density multiplied by its conductivity. The ferrite magnet strength multiplied by the speed of the rotating ferrites, then multiply that by the Torque. Take that figure, multiply it by the number of magnets and the distance from the surface of the magnetic event to give the magnetic event output. Uptake zone depth multiplied by the uptake mesh spun length multiplayer by the torque and number of magnets gives the magnetic event uptake. The magnetic event output and event uptake ant the output are all equivalent to each other.

7. Energy Increase Footprint:

Energy system total output minus the total input then add the gravity shield maintenance energy which gives a vector of total transferred energy over a period of time. This conserved reuptake energy is equivalent to the number of storage batteries multiplied by the held charge for each individual storage battery.

# Results & Analysis:

In order to protect the intellectual property rights of my myself and my research collaborators and financial backers at Innovative Asset Management certain data have been omitted which it is deemed could be detrimental or harmful to those individual rights. Therefore exact statistical and numerical figures, values and data which could otherwise lead to carbon copy replication of technologies, experiments and results have been omitted.

Therefore a significant amount of the results of this study are a meta-analysis of the original research study by Dr. Eugene Podkletnov of Moscow University into the 'gravity shielding' and dampening effects of high temperature superconducting ceramic disks, the evidenced gravity influencing effects of which were published and popularised in Nick Cook's book; The Hunt For Zeropoint: Inside The Classified World Of Antigravity Technology, (Cook, 2002). Podkletnov found that at liquid nitrogen cooled temperatures close to zero Kelvin and at extreme rotational speeds of approximately nine thousand revolutions per minute, enabled through the use of magnetic ferrite disk acceleration, that there was a calculated weight reduction, in all objects placed above the Gravity Shield, of approximately two to five percent.

This result is also consistent even when the objects of reduced weight are many floor strata in height above the operating device, in separate rooms, so as that the possibility of the weight reduction being a mechanical product of the rotational motion can be thought of as nullified by the presence of a physical barrier. This barrier works through absorbing the movement of any physical particles as they rise from the rotating ceramic devices disk surface, if exposed, by both the ceiling/floor which separates the weight

30

reduced object and the device and also by the physical casing of the device if the internal objects components are of an unexposed design variant as well.

The exact mechanism of action for the Gravity influencing effects of rapidly rotating (150rps) floatation magnet suspended, zeropoint cooled and vacuum tension sealed high-temperature superconductive ceramics, in the operation of the Gravity Shield, would at first initially be experimentally inconclusive due to the lack of clear-cut and defined replicated and verified data as to the specific recorded material and physical factors and pre-conditions for the manifestation of the effect. This is in part due to the highly advanced quantum mechanical nature of the operant causalities of the Podkletnov Gravity Shield and the relatively recent occurrence of the technological innovation and development of the Gravity Shield device on the timeline of the development of Gravity Physics.

All the more there is also a distinct lack of clear-cut and defined internationally verified statistical evidence for the justification of any theoretical causes and solid evidenced observed recorded effects of those suspected causes. This is also conflated and enshrouded by a general lack of any real documented and unified theoretical deductions, analysis, synthesis or conclusions put forward or academically published in the emerging area.

Also with the lack of any concordant experimental observations and supportive methodological falsifications of similar evidenced experimental effects of any other technological innovations and developments within similar areas of the developing Quantum Antigravity paradigm also contributes to the academic lack of phenomenological and theoretical definition and edifice of knowledge as to the existence of superconductive gravity effects which then makes difficult the job of discerning the quantum and potentially hyper-spatial phenomenal nature of the dimensional force and dimension we know as Gravity.

# Discussion:

According to quantum mechanics and superstring theory (Hawking, S., 1989; Hawking, 2001; Hawking, 2008) all hadronic matter, at its most basic fundamental level level of manifestation, consists of closed-ended vibrating heteroic strings, or loops of condensed and confined determinate finite energy packets, or quanta. These quanta inter-associate and vibrate in multiple dimensions at once, (Mulvey, J.H., 1981). I shall now discuss some of the main theoretical concepts that I shall then schematise into the paradigmatic perspective I call Hyper-particle Quantum Gravitation, beginning with zero-point energy.

If you take all the air out of a given space, then you create a vacuum, but still within this space there is a residual tensile energy left, vacuum energy: or zero point energy as it is also known. The Zero point energy within any typical massless vacuum is accordingly therefore not a true vacuum like the infinite abyss of the macro-cosmic void which manifested in the universes conception, before Hubble's 'Big Bang', but is a relative gas mass reduction induced vacuum. This unexplained anomalous residual background energy, which is detectable in residual vacuums, as a part of the fundamental nature of reality, independently of matter, is one of the foundations of the hyperspatial essence of Hyper particle Gravitation Theory.

The task for scientists now is to define what this residual energies essential nature is, then explain the mechanisms of action at work and then make suggestions of how we can influence the causation of its fundamental interactions; in order to bring coherency to its essential nature to make it a potential harnessed future energy resource. The electrical engineer M. King in his book on Zero-point energy and advanced electrical engineering in

2002, through deduction from observations of high energy plasmoid manifestation coherencies, which defines this background sea of vacuum energy as a 'quantum foam' or 'phenelum' of a residual manifest essence of metaphysical energy causation. High energy Plasmoid coherencies in which there exists, the spontaneous manifestation and then recession of interconnected sub-quantum white holes and black holes, in a multidimensionally stratified and internally coherent, yet superficially chaotic, unstable and fluctuating state of continuous multidimensional quanta virtual particle energy packet interaction.

In my 2010 book: Dawn Of The Neo-Modern, I speculated that this rapidly oscillating fluctuating invisible sea of electro-tensile energy may potentially cohere if an electromagnetic conductor were to wobble, oscillate and vibrate a computer enabled alternating current at the right frequencies. If done correctly this could harmonise the electromagnet with the zero point field essences frequency pattern, transforming it from a state of spontaneous fluctuation, into a state of harmonic resonance with the electromagnet, causing an exchange of energy causation from the vacuum 'phenelum' to the    field      of      the electromagnet, through the energy cohering and condensing effect of the alternating current wave pattern, (Garfoot, 2010). The next concept I shall discuss is that of an electroconductive luminaferious ether, of the existence of a multidimensional hyperspatial unifying metaphysical essence to the fundamental nature of all manifest matter and reality.

A Photon, or light quanta, is an innately massless fast moving quantum energy packet, which according to the 'double slit' experiment, displays the properties of both waves and particles. Most interestingly quantum energy packets of light also behave differently whether being recorded whilst observed by a human being  or unobserved being recorded by a computer; which raises philosophical questions about the dimensional essence of thought and consciousness being manifest in the fundamental nature of matter,.

This is because the Photon contains different amalgamated aspects of the combined essential natures of its manifest existence occurring in multiple dimensions all at once. The photons multidimensional nature explains how its path can be altered by a field of gravity, yet also have no mass so should according to logic and reason should be unaffected by the quantitative mass-weight causalities of the mechanisms of action of gravity.

A Photon is fundamentally a one dimensional point of individuation, a two dimensional electromagnetic wave which can bend around corners, a three dimensional particle which can bounce off physical objects, in its motion through space also means it exists in the fourth dimension as well; time. In order to exist as a wave though requires that it has a fundamental electromagnetic conductive medium of transmission, or Ether. An Ether in which the massless quantum energy-packet vibrating wavelength is a disturbance, or distortion, as it travels through the void of interstellar space.

So if there is the need for an electromagnetically conductive fifth dimensional ether in order to explain the behaviour and nature of photonic light, then surely we should consider the energy field nature of gravity and time in a similar way as well. Perhaps through having a degree of influencing causation upon each other, having an Ether of wave-particle conductive multidimensional weave to the essential nature of their manifest causation upon each other.

It could be considered here that there is indeed the need for a gravity, light & time conductive all pervading hyperspatial ether within the paradigms of Superstring theory, Gravity, Quantum Mechanics, Photonics, Zero point Energy and Physics as a whole. This is in order to explain the symbiotic nature of the path of light and electromagnetic radiation, the passage of Time and warping effects upon both of Gravity according to Einstein's Relativity, (Carlip, S., 2019; Einstein, A., 1916; Einstein, A., 1922)

This ether warping capacity upon the flow of time through the altering of the path of Photons by strong fields of gravity such of those of black holes could be explained by the notion that there must be some degree of fundamental flex to the interwoven metaphysical unification of the dimensional properties of electroconductive hyperspace. Thus suggesting a unifying mathematical simplification that this could bring to our understanding of the laws of nature through the inclusion of these higher dimensions of reality (Kaku, 1990).

Therefore it is not just that the nature of matter, but more, that part of the fundamental nature of reality that we all exist in ten dimensions, as both physical matter and multidimensional energy wavelength vibrations that are emitted out. So, if point particles are multidimensional vibrating strings attached through electron cloud tension to an all pervading hyperspace oscillating out their energies at a fundamental level, then in as much as atomic nuclei exist, their electron cloud energy wave emissions exist also, and the two are unified.

Gravity is a property of matter, an energy caused by multidimensional superstring vibration, radiating out into universe as an attraction manifesting property of a universal and fundamental force. The medium of its quantum energy packet transmission, the Gravity Ether permeates and penetrates all physical reality, and the Graviton energy packet that moves through this medium, does so largely according to the already established laws of Quantum Mechanics. The Graviton as an individuated particular of causality has its effects upon other atoms through the exchange of virtual particle energy interactions with a strength of effect and direction of concentration according to the Gravitons range or proximity, and direction of locational focus.

It is a possibility for all the dynamic energy interactions to be the product of the tensile causality of superstring vibrations contained within and transmitted across the multidimensional unifying medium of a zero-point energy conductive hyperspatial ether. An ether which exists as an agent of particle-wave transmission and virtual energy and particle interactions, gravity and and mass being each both cause and effect. The cause as the source of gravitonic energy, in one respect being the cause of the effect, but also being subject to the effects of gravity as the target of the movement of the virtual particle energies which have similar gravity causal natures of quantum interaction.

The medium that gravity vibrations move through I call hyperspatial gravity conductive ether, it conducts the Graviton energy packet as a wave/particle through time and space to an extent which is still currently only definable in divisibility and range, like its dimensional cousins time and space as: 'an infinite given magnitude' (Kant, I., 2021). What we have to consider here is that the graviton is a massless particle-wave, in some respects akin to its five dimensional cousin, the Photon, but due to the simplification of the laws of nature; it can tunnel infinitely. Which would allow for the forces weakness of attractive energy range through its unbound limitless infinite magnitude, and its unhindered indivisible penetration of all physical matter. Gravity exists not just as a force of nature and as a fundamental energy of the universe, but also as also a dimensional essence of reality in its own right.

Gravity is the current day manifestation of the fundamental essence which once bound the proton-universal singularity together as an infinitely small and dense point of one dimensional individuation, just before the other three expansive dimensional forces exploded into existence at the moment of the universes manifestation. This is the fundamental starting point of creation, when four dimensions expanded and the six remaining dimensional essences of the universe collapsed forming the multidimensional energy conductive tensile medium of the hyperspatial ether at the point when the universe spontaneously came into existence.

From these composite theories I have briefly outlined thus far I shall formulate a synthesis from their aspects which are relevant and in agreement with each other. From which I shall briefly sketch a new composite of theoretical concepts and visualisations, which will explain all of the observed evidence and align the paradigm with the pre-existing perspectives into The Infinite Energy Theorem and Hyper particle Gravitation Paradigm.

If gravity waves are through the simplification of the laws of nature through unifying higher dimensions massless and can tunnel infinitely, or without limit, then they should also logically, as a part of their multidimensionally transcendental essential nature, have infinite penetration into and through matter, as a property, as well. This allows the transcendental gluon-graviton matter-energy vibration ejection carrier wave transfer of energy, as a metrically tensile mechanism of action, in the form of the fundamental attraction of all physical objects of mass to all other physical objects of mass, to occur, through the effect of gravity.

This is not just from the behavioural nature of the Graviton, but also is because of the causation of the latent gravity energy charge ejection and gravity conductive properties of the Gluon matrix; which composes the fundamental constituent of a particles mass and binds the hadronic nucleus of an atom together at its subatomic core. The Infinite nature of the magnitude of Gravity, and the tunnelling range of the Graviton, could be considered the fundamental basis, for the classification of Gravity, as a hyper-spatial dimension. This is as it has the essential property, of having an infinite quality, to its fundamental nature, in the form of the extent of its magnitude, as a fundamental dimensional aspect of reality, to the manifestation of matter, and the nature of physical existence.

This is because gravity is not only a hyperspatial quality of matter and reality, but the graviton itself is a six dimensional particle, whose laws of manifest essential nature, contain a further fundamental dimensional simplifications. Like the five dimensional massless essence of Photonic electromagnetic ether extended to make it a six dimensional tensile ether whose Gravitonic energy conductive quanta have further meta-spatial and meta-temporal properties in addition to the nature of their causation.

The multidimensional meta-temporal and meta-spatial essence to a Hyper particle in its movement and the path of its motion ranges from the point of origin to its destination. From beyond the vast void which separates the universe from its omni-dimensional reality containing metaphysical barrier through which gravity can penetrate. Passing through the fundamental binding tension of the gravity energy laden Higg's Boson embedded in the Gluon matrix of every atom in the universe, picking up tensile gravity energy as it travels through it as a concentric energy pulse. As the Gravity pulse passes through the gluon matrix of nearby atoms it ejects gravity energy from the zero point tension of the Gluon binding of the core of the atom similar to how photons eject electrons according to the impact resonance frequency and wavelength of the photon that ejects it into motion.

When the Hyper particle Gravity pulse reaches its end point of destination inside the nucleus of the atom that it is tethered to it causes a spontaneous gravity spasm in the tension of the Gluon Matrix relative to the proximity of matter (as the transmitted gravity charge reduces with range) and its direction (as all gravity charge, is transmitted from all sources as one constant pulse, due to its meta-temporal nature and infinite concentric motion). The question here is, what is it which lies between the two resonances of the Gluonic binding tension energy matrix and the Gravitonic hyper particle energy pulse? The answer to this question, quite nicely defines, the theoretical foundations, of our understanding, of the influencing causalities, upon the mechanisms of action of Gravity, which the gravity shield has; and the answer is the electron cloud quantum energy field.

Could the electromagnetic resonance of the superconductive disk, with its metallic bond enabled degree of ceramic contained electro-acoustic resonance, harmonised between electrons, in the ceramics superconductive overlapping shell structure, interfere with the graviton-glueonic energy exchange? Would this have an energy dampening effect, reducing the strength of Gravity upon all objects placed above it? I will now attempt to justify the unique property causations, of the gravity inhibiting and dampening effect of high temperature superconductors, at the quantum level, explaining exactly how a superconductive rotating disk can interfere with the normal operation of a dominant field of gravity.

The theorem known as the Heisenberg Uncertainty Principle the dynamics of the point particle localised individuation of substance and electro-acoustic resonance in ceramics and metals is essential to our understanding of the causation of the effect of superconductor gravity dampening at the quantum level. In the normal functioning of the conductive nature of the electron shells, in the form of metallic bonds, because of the close proximity of the shells to each other, so that they overlap and interlock, there is a delocalised free range of the movement of electrons through the material as they jump and hop from one atoms electron shells to another.

As the electrons fly around in the metal or ceramic they collide into the occasional nucleus of an atom, creating thermal energy but usually bending round according to the shape carved out of the electron cloud/shell interlocking composite of the metallic bonds. As they do so they lose energy to resistance and cause the atom to wobble, the result of this is heat.

The orbit of an electron around a proton is perpetually maintained by its attraction to the core of the atom unless its energy threshold increases and the electron moves to an outer shell in the spectral line electron cloud configuration and the phase state of the metal changes, eventually becoming an atomic plasma as all the electrons have escaped their individual Fraunhofer lines. When a metal is cooled however, the effective size of the nucleus of the metal shrink and reduce their resistance causation and electron flow influencing effects upon the electricity which is conducted through it.

`Once the material goes superconductive then all of the electrons lose the attractive energy which maintains their continued movement and the intrinsic chaos to their motion ends as the vibrations between them harmonise becoming coherent as the system of energy evolves from a state of chaos and flux, towards a state of order.

As the electrons harmonise, the connective surface tension of the electrons in their new superconductive state increases according to the Heisenberg Uncertainty Principle of locality. This states that the more we can tell you about the variable of the speed of a particle, the less we can tell you about the variable of it's location. The result of this is a motion blur elongated smudge of the spherical particle in a snapshot of that particles location and motion at any given time, (Heisenberg, W., 1930; Heisenberg, W., 1979)

So, if we slow down the motion of the freely moving non-local electron particles in a charged low temperature superconductor so that they come to an absolute stop. Then through removing the motion influencing and electrically resistive causation of the oppositely charged protons, in the nucleus, which would otherwise attract the electrons and maintain their motion in the overlapping electron shells/clouds of the metallic/ceramic bonded compound.

Then the elongated distortion of the point particles will, eventually, begin where the motion ends, turn into a circle; which you can then plot on a hyper particle sphere. Their location can be accurately known, at the point of superconductivity, as the electron stops moving and the edge of the particles manifest boundary of solid individuation increases in the solidified density of its boundaries. Now that it's edge has re-cohered from a low level of edge solidification and high degree of energetic chaos into an order of high level solidification and low degree of chaos.

As has already been discussed, the multidimensional nature of the movement of a graviton hyper-particle is essentially concentric as it moves through nearby matter ejecting quantum gravity charge from nearby gluon enmeshed gravity conductive Higgs Boson. As it moves inwards towards the Gluon matrix of the particle with which it is associated and fundamentally tethered.

Here is the explanation for why it is that rotational motion to such extreme levels of speed of the rotating ceramic superconductor disk is so essential to the production of the gravity dampening effects of the gravity shield. The electrons, harmonised through their electro-acoustic vibration into a state of unity, with definite non-chaotic individuation to their essential nature, are rotated, and once more according to Heisenberg's theory, become an elongated motion blur, though this time with a solid defined edge of physical individuation, as opposed to a chaotic and transient edge of individuation if the leptonic particle is in motion in metallic or ceramic bonds.

As the ceramic disk spins faster and faster the length of the Heisenberg motion blur of the electron suspended in electro-acoustic resonance in the metal motionless but under the rotational movement of the disk becomes a blur through the intense rotational motion. Eventually, bent in a circular curve by the disks motion, the circular overlap of the elongated motion blur of electrons begins to stretch far enough as to occupy all points in its particular locality in the rotating disk.

As a particle now occupying a circle of manifest locality effect causation through satisfying the criteria for both quantum and hyper-particle paradigms of gravity physics together, which allows for the alignment of causation into the creation of the antigravity hyper particle gravitation effect. This is because the hardened shell of the point particle individuation of the electron corresponding to the transient embedded nature of the graviton as a physical object with determined spatial physical bounds and causalities of finitude that intersect and overlap as a composite mathematical fractal structure. This means the individuation hardened lepton has the impact harmonics, motion, and hyper-spatial alignment necessary to deflect the graviton in its concentric path of motion through the gravity shield ceramic and cause the gravity dampening effect to the extent it does.

This is through the manifestation causality of the hyperspatial alignment of the intrinsic multidimensional nature of the knit between the three dimensions of space and seven dimensions of energy according to the Ancient Pythagorean value of Pi. As the only infinite decimal natural value and the figure usually associated with the calculation of circular shapes and the manifestation of spheres as a intrinsic natural configuration of matter, it is only natural its value would have importance in its fundamental embedded manifest dynamics of the hyperspatial multidimensional nature of the universe; thus ends this discussion.

# Conclusion:

Because the same gravity effect input into the system by the Gravity Shield can create varying amounts of torque transferable energy using the same amount of input energy, this creates an exponential ratio of energy gain output to energy input defined as the energy increase footprint. Through the conservation of energy it is possible to create an energy accumulation power system where not only are there no losses of the initial spark of energy, because of the nature of kinetic energy itself.

Energy in the form of momentum, the energy from the flywheel is through floatation magnet suspension and sealed vacuum motion reinput back into the system after the output has been drawn out of the system, creating a feedback loop of an ever-increasing accumulation of potential Gravitonic energy, as a power source in the form of kinetic and electric energy. It is my belief that through creating electro-acoustic effects within superconductors we can effectively sandwich an interference layer of harmonised Electricity coherent at the energy vibration levels of the Gravito-glueonic energy transfer between the Gravitonic and Gluonic layers of quantum reality. Therefore creating resistance to the transfer of Gravitonic and Gluonic energies reducing the effect of gravity of any object placed above it (Gillies, J., 2018). This explains the effect upon gravity of the Gravity Shield founding the basis for Gravitoelectric Power; effectively, utterly and completely solving the world, and human races, need for energy as long as it's civilisation exists.

# Bibliography & References:

Aristotle, (2008) Physics, Oxford University Press: Oxford.

Carlip, S., (2019) General Relativity: A Concise Introduction, Oxford University Press: Oxford.

Cook, N., (2002) The Hunt For Zero-Point: Inside The Classified World Of Antigravity Technology, Palgrave: London.

Einstein, A., (1916) Relativity: The Special & The General Theory, Penguin Classics: London.

Einstein, A., (1922) The Meaning Of Relativity, Princeton University Press: New York.

Garfoot, A.P., (2010) Dawn of The Neo-Modern: Art, Humanism & The Meme. Lulu: London.

Garfoot, A.P., (2012) Faster Than Light Travel & How To Destroy A Dwarf Star, Lulu: London.

Gillies, J., (2018) CERN and the Higgs Boson: The Global Quest for the Building Blocks of Reality, Icon: New York.

Hawking, S., (1989) A Brief History of Time, Bantam Books: London.

Hawking, S., (2001) The Universe In A Nutshell, Bantam: London.

Hawking, S., (2008) The Theory of Everything: The Origin and Fate of the Universe, Jaico: London.

Heisenberg, W., (1930) Physical Principles of the Quantum Theory, Dover: New York.

Heisenberg, W., (1979) Philosophical Problems of Quantum Physics, Oxford University Press: New York

Kaku, M., (1995) Hyperspace: A Scientific Odyssey Through Parallel Universes, Timewarps & The Tenth Dimension, Oxford University Press: Oxford.

Kant, I., (2021) Critique of Pure Reason, Oxford University Press: Oxford.

King, M., (2002) Tapping The Zero-Point Energy: How

"Free Energy" & "Anti-Gravity" Might Be Possible With Today's Physics, Adventures Unlimited Press: New York.

Kuhn, T., (1962) The Structure Of Scientific Revolutions, University of Chicago Press: Chicago.

Mulvey, J.H., (1981) The Nature Of Matter, Oxford University Press: Oxford.

Newton, I., (2018 [1726]) Newton's Principa: The Mathematical Principles of Natural Philosophy, Forgotten Books: London.

Sachs, J., (1995) Aristotle's Physics: A Guided Study, Rutgers Press: New Brunswick.

# Self Evolving Energy Dynamic Flowchart

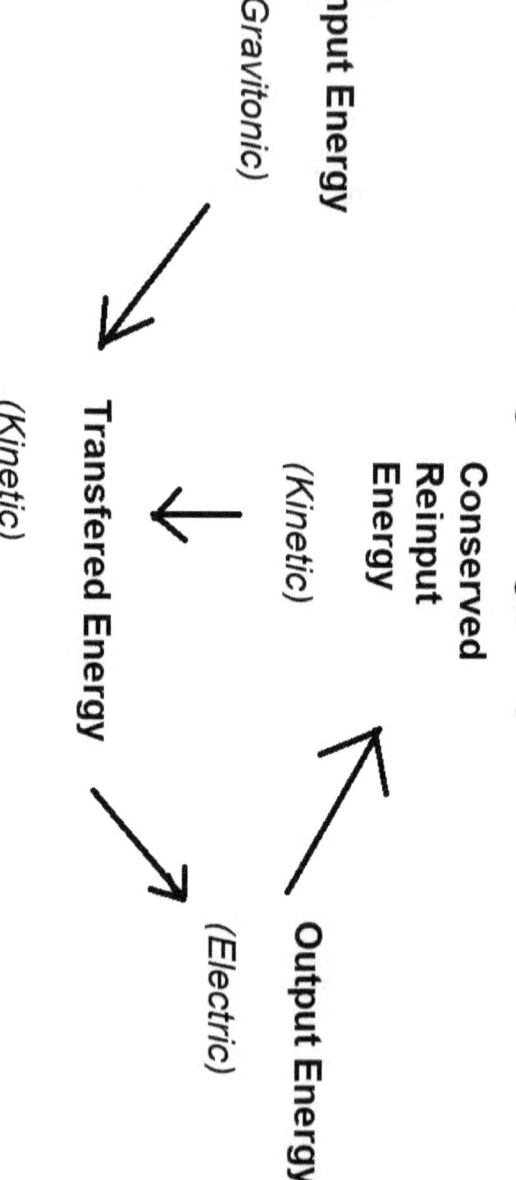

**Input Energy**
*(Gravitonic)*

**Conserved Reinput Energy**
*(Kinetic)*

**Transfered Energy**
*(Kinetic)*

**Output Energy**
*(Electric)*

# Energy Loss Conservation Flowchart

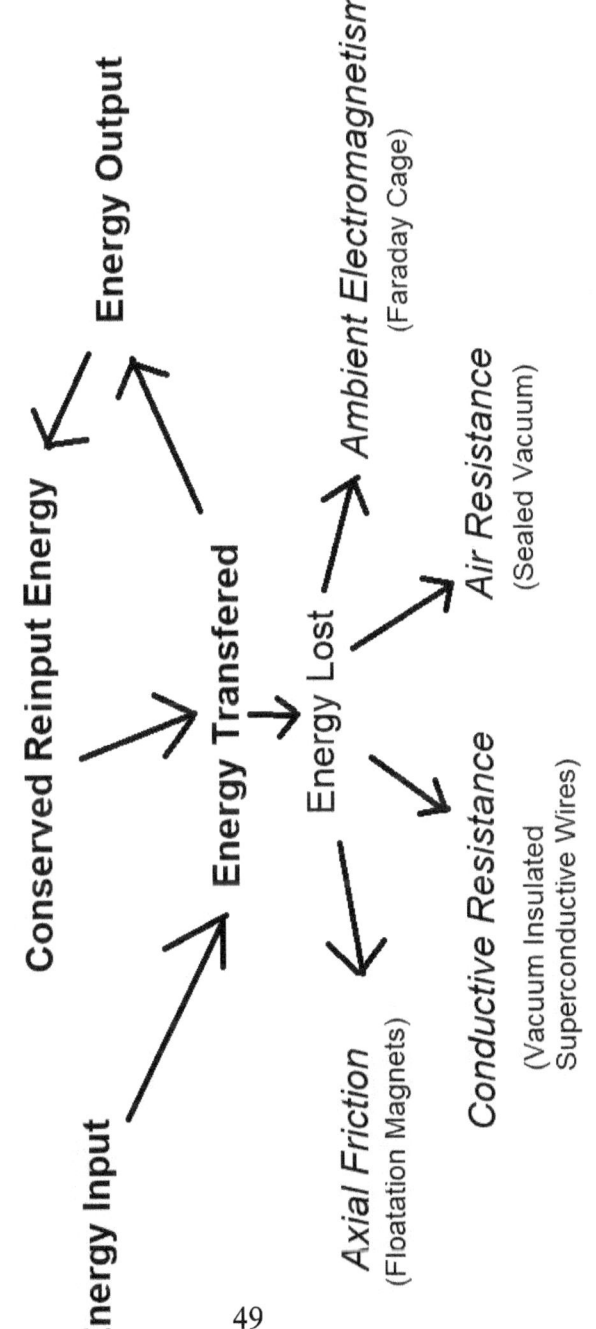

**Energy Output**

**Conserved Reinput Energy**

**Energy Transfered**

**Energy Input**

Energy Lost

*Ambient Electromagnetism*
(Faraday Cage)

*Air Resistance*
(Sealed Vacuum)

*Axial Friction*
(Floatation Magnets)

*Conductive Resistance*
(Vacuum Insulated
Superconductive Wires)

49

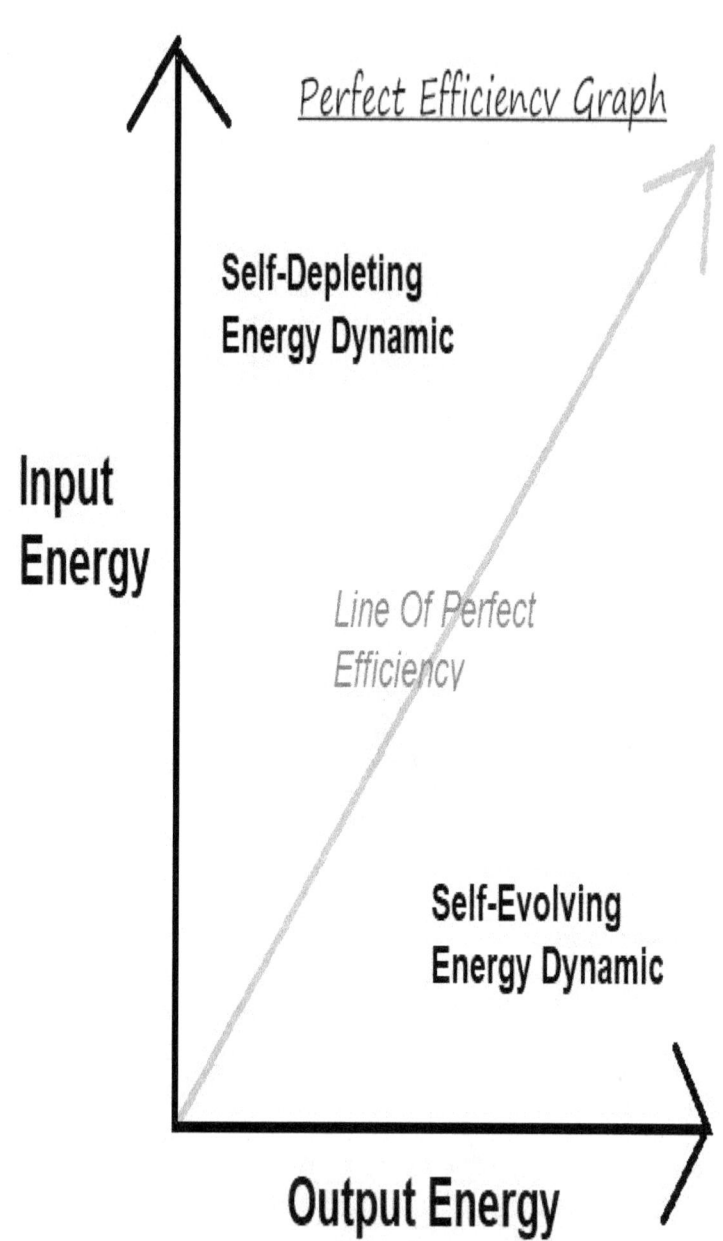

Perfect Efficiency Graph

Self-Depleting
Energy Dynamic

Input
Energy

Line Of Perfect
Efficiency

Self-Evolving
Energy Dynamic

Output Energy

# *Output Lifetime Extention Graph*

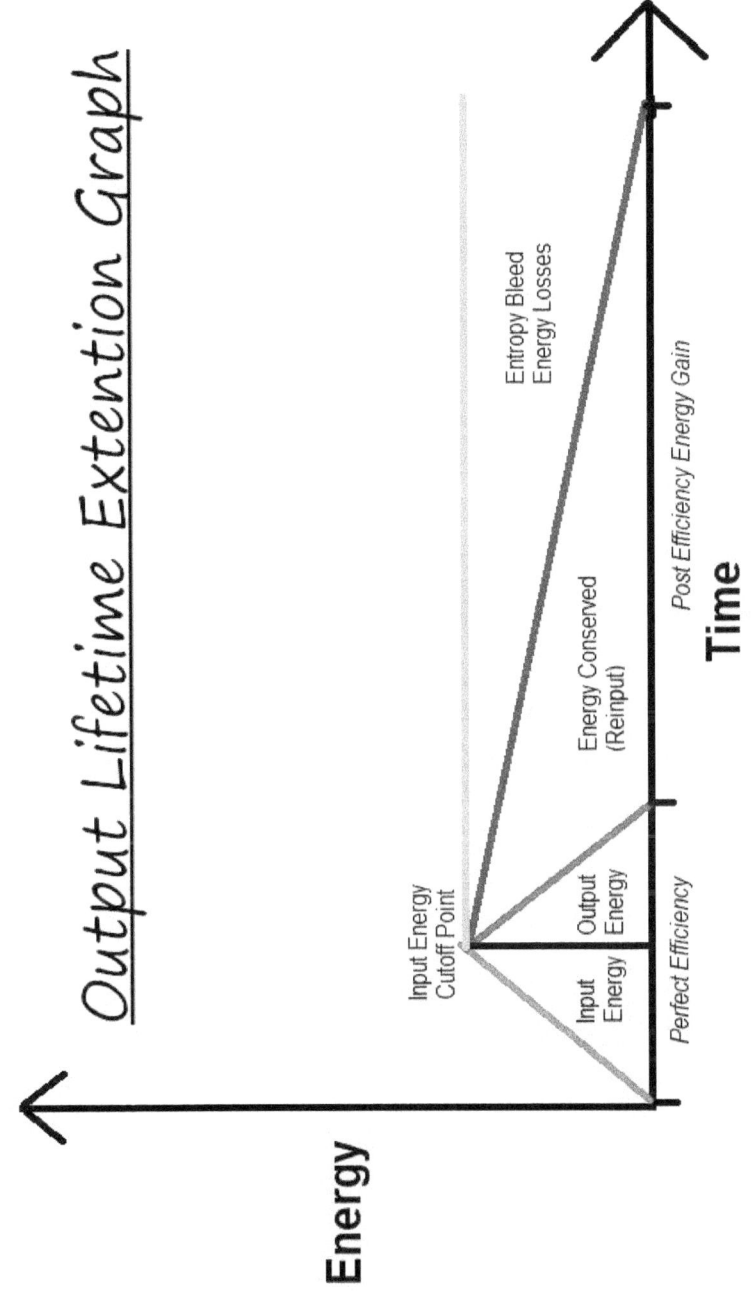

**Energy**

**Time**

Input Energy
Cutoff Point

Output
Energy

Input
Energy

*Perfect Efficiency*

Entropy Bleed
Energy Losses

Energy Conserved
(Reinput)

*Post Efficiency Energy Gain*

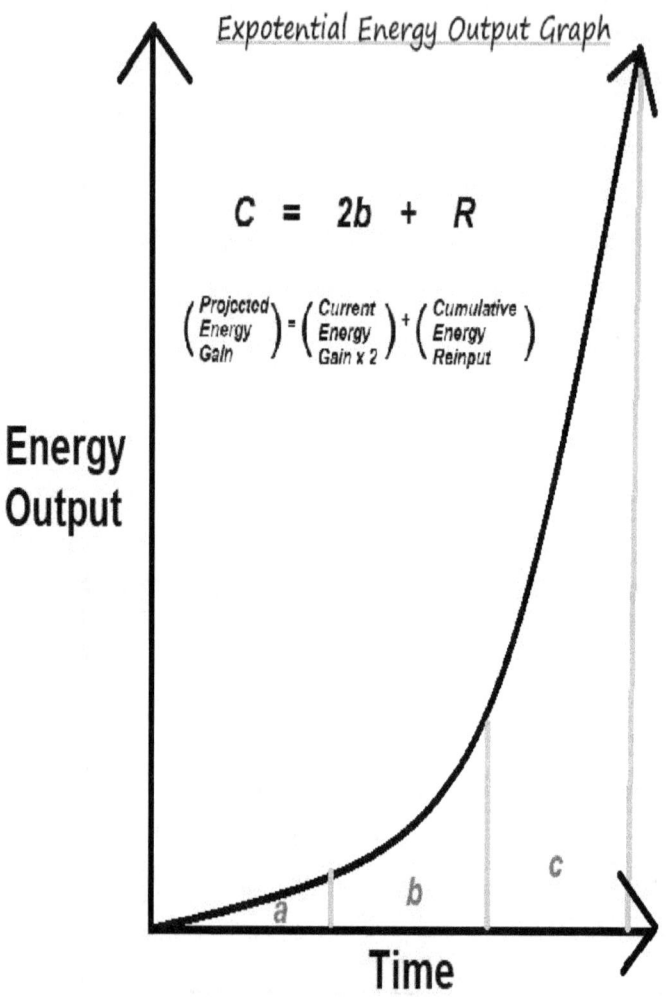

Expotential Energy Output Graph

$$C \;=\; 2b \;+\; R$$

$$\left( \begin{matrix} Projected \\ Energy \\ Gain \end{matrix} \right) = \left( \begin{matrix} Current \\ Energy \\ Gain \; x \; 2 \end{matrix} \right) + \left( \begin{matrix} Cumulative \\ Energy \\ Reinput \end{matrix} \right)$$

**Energy Output**

*a*   *b*   *c*

**Time**

# Output & Initiation Energy Graph

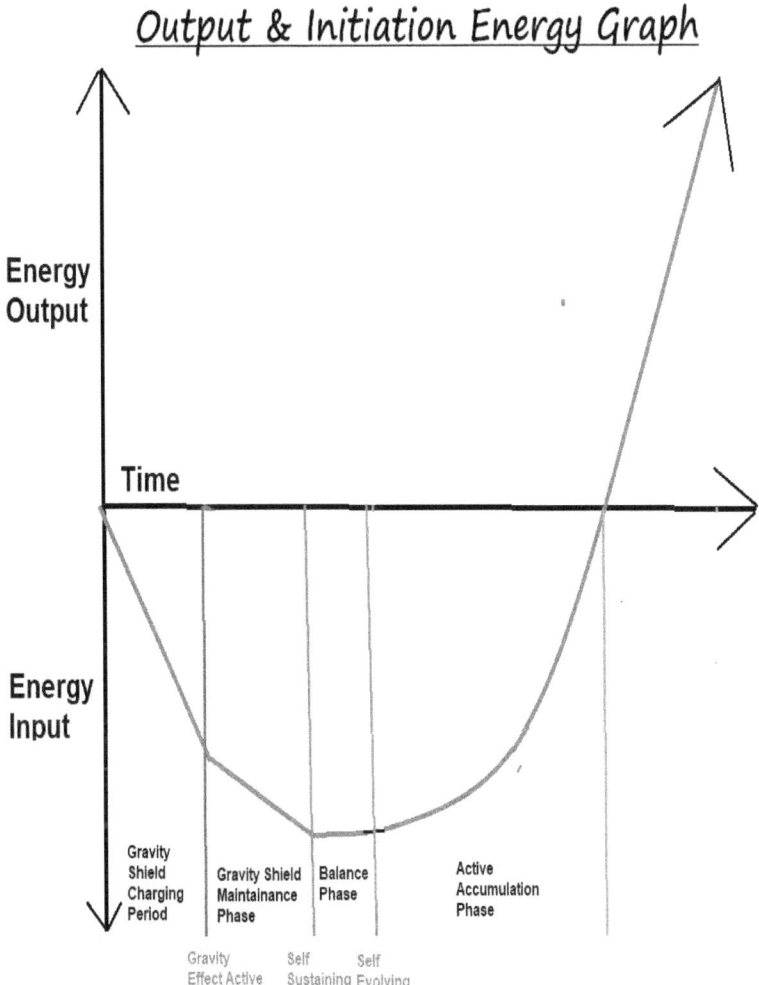

# Weight & Momentum Equation

$$\mathbf{W} = \left( \left( \mathbf{+} + \mathbf{\boxminus} + \mathbf{\nearrow} \right) + \left( \diamondsuit \times \mathbf{M}^n \right) \right) = \left( \wedge = \backslash = \mathbf{\sqsubset} \right)$$

$\begin{pmatrix} \text{Total} \\ \text{Weight} \end{pmatrix}$ $\begin{pmatrix} \text{Axial} \\ \text{Beam} \\ \text{Weight} \end{pmatrix} \begin{pmatrix} \text{Uptake} \\ \text{Array} \\ \text{Weight} \end{pmatrix} \begin{pmatrix} \text{Flywheel} \\ \text{Weight} \end{pmatrix}$ $\begin{pmatrix} \text{Weight} \\ \text{Of Array} \\ \text{Magnet} \end{pmatrix} \begin{pmatrix} \text{Number} \\ \text{Of} \\ \text{Magnets} \end{pmatrix}$ $\begin{pmatrix} \text{Floatation} \\ \text{Magnet} \\ \text{Strength} \end{pmatrix} \begin{pmatrix} \text{Load} \\ \text{Bearing} \\ \text{Strength} \end{pmatrix} \begin{pmatrix} \text{Number} \\ \text{Of} \\ \text{Magnets} \end{pmatrix}$

# Gravity/Torque Transfer Equation

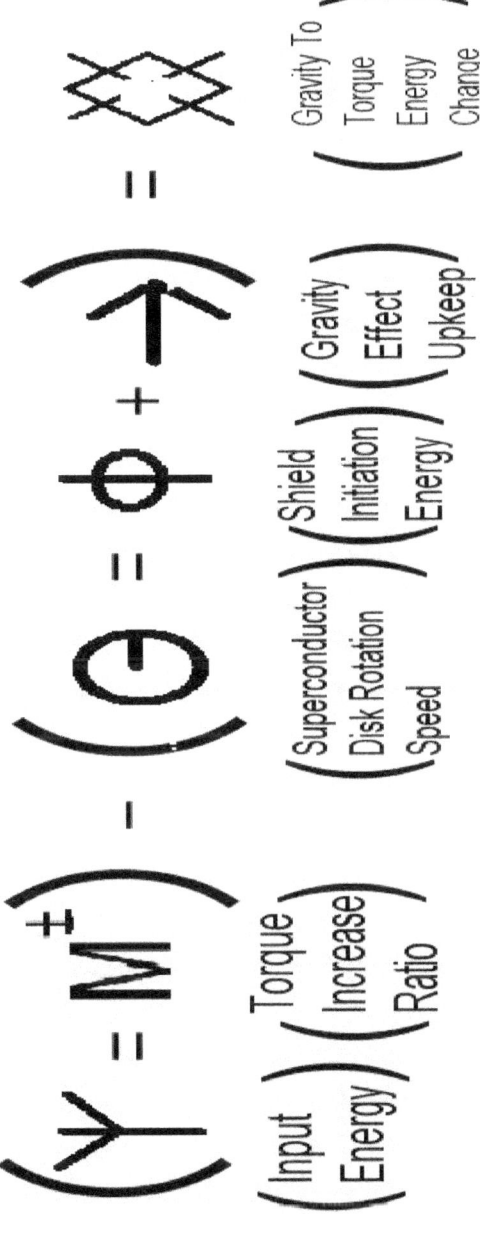

$$\left( \curlyvee = M^{\pm} \right) - \left( \left( \ominus = \phi \right) + \uparrow \right) = \lozenge\lozenge$$

$$\left( \begin{array}{c} \text{Input} \\ \text{Energy} \end{array} \right) \left( \begin{array}{c} \text{Torque} \\ \text{Increase} \\ \text{Ratio} \end{array} \right) - \left( \left( \begin{array}{c} \text{Superconductor} \\ \text{Disk Rotation} \\ \text{Speed} \end{array} \right) \left( \begin{array}{c} \text{Shield} \\ \text{Initiation} \\ \text{Energy} \end{array} \right) + \left( \begin{array}{c} \text{Gravity} \\ \text{Effect} \\ \text{Upkeep} \end{array} \right) \right) = \left( \begin{array}{c} \text{Gravity To} \\ \text{Torque} \\ \text{Energy} \\ \text{Change} \end{array} \right)$$

55

## Conserved Energy Feedback Loop Equation

$$\left( \text{ᚻ} \times \triangle \times \uparrow = M \right) + \left( \triangle \times t \right)$$

$$\begin{pmatrix} \text{Flywheel} \\ \text{Rotation} \end{pmatrix} \begin{pmatrix} \text{Rotation} \\ \text{Rate} \end{pmatrix} \begin{pmatrix} \text{Weight Of} \\ \text{Flywheel} \end{pmatrix} \begin{pmatrix} \text{Torque} \end{pmatrix} \qquad \begin{pmatrix} \text{Accumulated} \\ \text{Energy} \end{pmatrix} \begin{pmatrix} \text{Elapsed} \\ \text{Time} \end{pmatrix}$$

$$\left( \text{\#} - \ominus \right) = \left( \bowtie - \curlyvee \right)$$

$$\begin{pmatrix} \text{Energy} \\ \text{Dynamic} \end{pmatrix} \begin{pmatrix} \text{Total Energy} \\ \text{Accumulated} \end{pmatrix} \begin{pmatrix} \text{Total} \\ \text{Energy} \\ \text{Input} \end{pmatrix}$$

$$= \begin{pmatrix} \text{Cumulative} \\ \text{Torque} \\ \text{Increase} \end{pmatrix} \begin{pmatrix} \text{Energy} \\ \text{Reinput} \end{pmatrix}$$

# Energy Lost Through Entropy Bleed Equation

$$\left(\text{⚡} + \text{↯} + \text{◮}\right) + \left[\left[\left(G - \otimes + \text{⤬}\right) + \left(\underline{F} - F'\right) + \left(P - \underline{P}\right) \times t\right]\right] = E$$

(⚡) Shield Initiation Energy
(↯) Ferrite Charging Energy
(◮) Vacuum Initiation Energy
(G) Shield Air Resistance
(⊗) Shield Vacuum Initiation
(⤬) Effect Upkeep Energy
(F) Array Axial Friction
(F') Frictionless Floatation Ferrites
(P) Array Air Resistance
(P) Uptake Array Vacuum
(t) Elapsed Period Of Time
(E) Total Energy Losses

57

$$M^n \times ( \square \times \lozenge ) = ( \wedge \times \Pi ) = O$$

$M^n$
$\begin{pmatrix} \text{Torque x} \\ \text{Number Of} \\ \text{Magnets} \end{pmatrix}$

$\square$
$\begin{pmatrix} \text{Size Of} \\ \text{Charged} \\ \text{Ferrite} \end{pmatrix} \begin{pmatrix} \text{Ferrite} \\ \text{Field} \\ \text{Strength} \end{pmatrix}$

$\wedge \times \Pi$
$\begin{pmatrix} \text{Energy} \\ \text{Uptake} \\ \text{Yield} \end{pmatrix} \begin{pmatrix} \text{Number} \\ \text{Of Uptake} \\ \text{Zones} \end{pmatrix}$

$O$
$\begin{pmatrix} \text{Harnessed} \\ \text{Energy} \\ \text{Output} \end{pmatrix}$

# Energy Transfer Dynamics Equation

$$\left( \exists \times H = \mathscr{L}\!\!\uparrow \right)$$

$$\left( \begin{array}{c} \text{Uptake} \\ \text{Material} \\ \text{Density} \end{array} \right) \left( \begin{array}{c} \text{Uptake} \\ \text{Material} \\ \text{Conductivity} \end{array} \right) \left( \begin{array}{c} \text{Ferrite} \\ \text{Field} \\ \text{Strength} \end{array} \right)$$

$$\left( \mathscr{L}\!\!\uparrow^{i} \times \sqcap \times M^{n} \right)$$

$$\left( \begin{array}{c} \text{Ferrite} \\ \text{Field} \\ \text{Strength} \end{array} \right) \left( \begin{array}{c} \text{Ferrite} \\ \text{Rotation} \\ \text{Speed} \end{array} \right) \left( \begin{array}{c} \text{Torque x} \\ \text{Number} \\ \text{Of Magnets} \end{array} \right)$$

$$\overline{ \quad\quad\quad r^{\%} \quad\quad\quad }$$

$$\left( \begin{array}{c} \text{Magnetic} \\ \text{Event} \\ \text{Distance} \end{array} \right) = \ \ominus \ \left( \begin{array}{c} \text{Magnetic} \\ \text{Event} \\ \text{Output} \end{array} \right)$$

$$\left( d \times \bowtie \times M^{n} \right) = \curlyvee$$

$$\left( \begin{array}{c} \text{Uptake} \\ \text{Zone} \\ \text{Depth} \end{array} \right) \left( \begin{array}{c} \text{Uptake} \\ \text{Mesh Wire} \\ \text{Length} \end{array} \right) \left( \begin{array}{c} \text{Torque x} \\ \text{Number Of} \\ \text{Ferrites} \end{array} \right) = \left( \begin{array}{c} \text{Magnetic} \\ \text{Event} \\ \text{Uptake} \end{array} \right)$$

$$\left( \ominus \quad \curlyvee \quad \bigcirc \right)$$

$$\left( \begin{array}{c} \text{Magnetic} \\ \text{Event} \\ \text{Output} \end{array} \right) \left( \begin{array}{c} \text{Magnetic} \\ \text{Event} \\ \text{Uptake} \end{array} \right) \left( \begin{array}{c} \text{Harvested} \\ \text{Energy} \\ \text{Yeild} \end{array} \right)$$

$$\left( O - I - \underline{\triangle} = \right) + \left( \bowtie = \parallel - \parallel \times \not\parallel \right)$$

$\left(\begin{array}{c}\text{Generator}\\\text{Total Output}\end{array}\right) \left(\begin{array}{c}\text{Generator}\\\text{Total Input}\end{array}\right) \left(\begin{array}{c}\text{Gravity}\\\text{Shield}\\\text{Input}\end{array}\right) \left(\begin{array}{c}\text{Total Energy}\\\text{Transfered}\end{array}\right)$

$\left(\begin{array}{c}\text{Conserved}\\\text{Reuptake}\\\text{Energy}\end{array}\right) \left(\begin{array}{c}\text{Number}\\\text{Of}\\\text{Batteries}\end{array}\right) \left(\begin{array}{c}\text{Stored}\\\text{Energy In}\\\text{Battery}\end{array}\right)$

60

# A. System Pan Overview

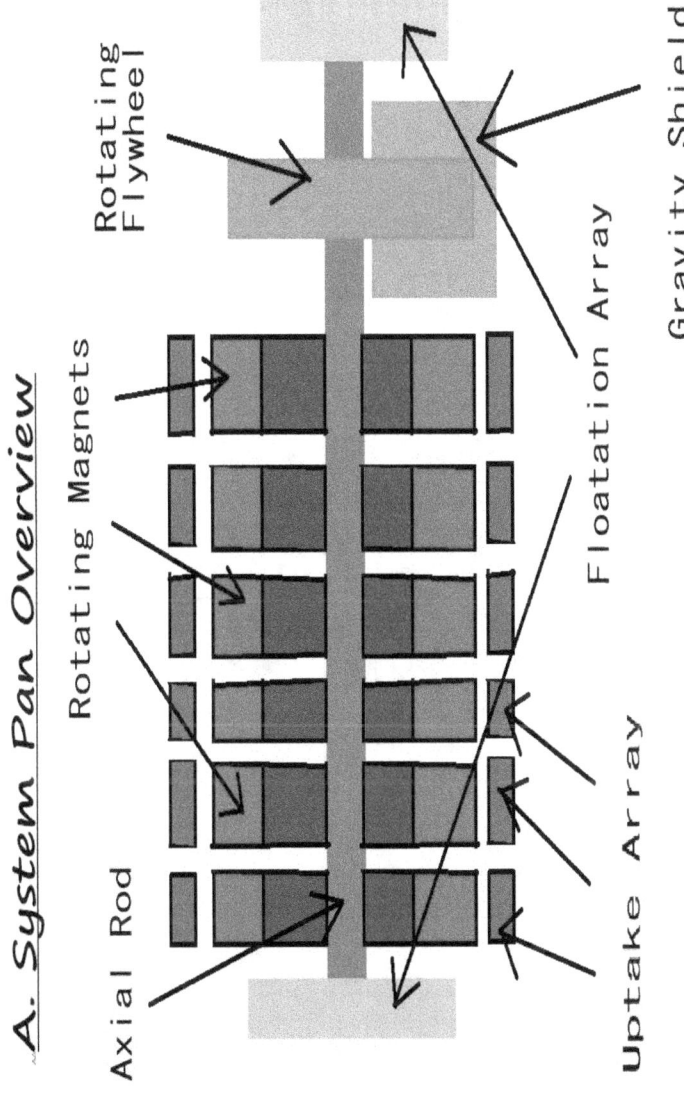

Rotating
Flywheel

Rotating Magnets

Axial Rod

Floatation Array

Gravity Shield

Uptake Array

# B. Floatation Magnet Array

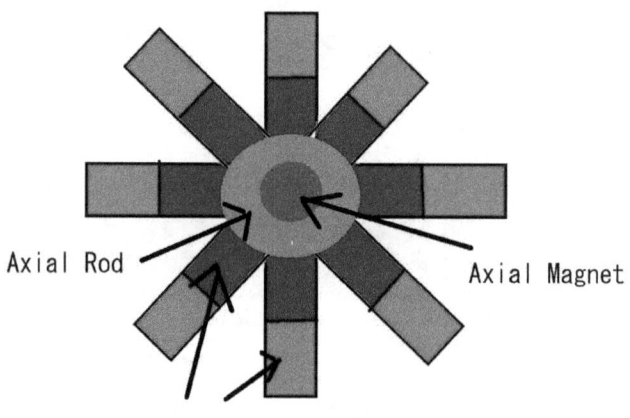

Axial Rod

Axial Magnet

Floatation Ring
Magnet

## C. Rotational Flywheel Cross-Section

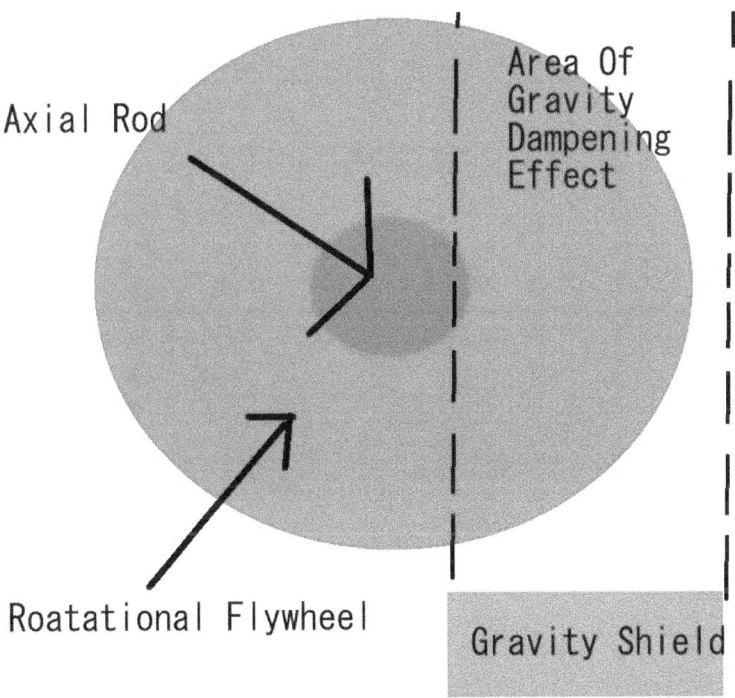

Axial Rod

Area Of
Gravity
Dampening
Effect

Roatational Flywheel

Gravity Shield

## D. Uptake Array Cross Section

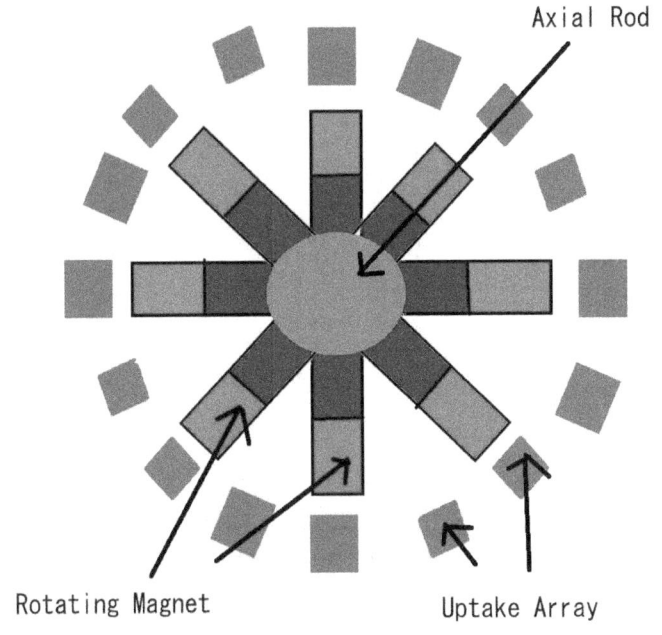

Axial Rod

Rotating Magnet

Uptake Array

65

# E. Gravity Shield Cross-Section

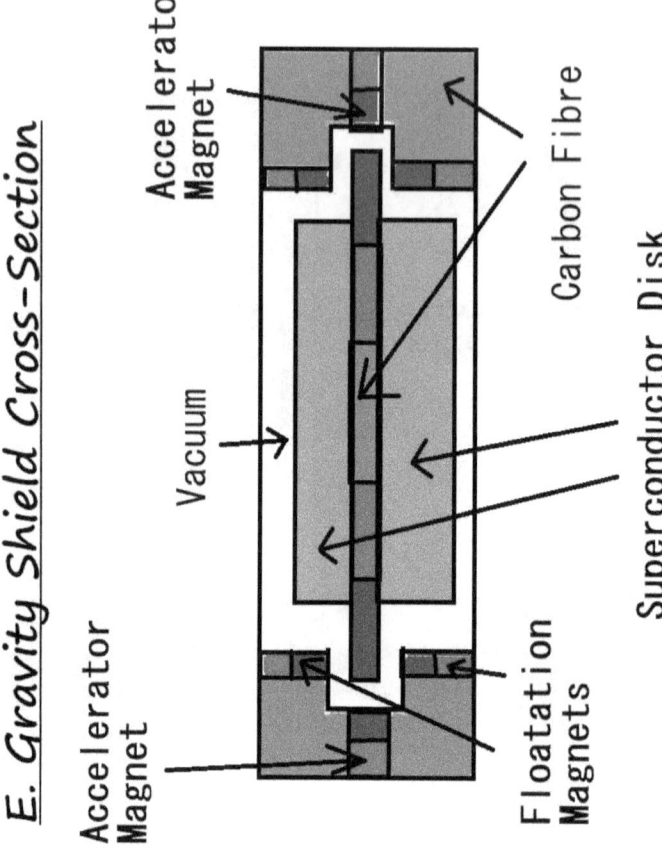

Accelerator Magnet

Accelerator Magnet

Vacuum

Carbon Fibre

Superconductor Disk

Floatation Magnets